George Herbert Taylor

**The movement cure in every chronic disease**

a summary of its principles, processes, and results

George Herbert Taylor

**The movement cure in every chronic disease**
*a summary of its principles, processes, and results*

ISBN/EAN: 9783337729295

Printed in Europe, USA, Canada, Australia, Japan

Cover: Foto ©berggeist007 / pixelio.de

More available books at **www.hansebooks.com**

THE

# MOVEMENT-CURE

IN

# EVERY CHRONIC DISEASE:

## A SUMMARY

OF ITS

### PRINCIPLES, PROCESSES AND RESULTS.

**BY GEO. H. TAYLOR, M. D.,**

Author of An Exposition of the Movement-Cure ; Physician of the Remedial Hygienic Institute of the City of New York.

𝔑𝔢𝔴 𝔜𝔬𝔯𝔨:

ROBERT LARTER, PRINTER, 610 SIXTH AVE.

—

1861.

# THE MOVEMENT-CURE.

## GENERAL PRINCIPLES.

### INSUFFICIENCY OF THE CURRENT PRACTICE OF MEDICINE.

To the careful observer, it must be quite apparent that much has yet to be gained in the field of Medical Practice, before it can boast of approximating in any marked degree toward perfection.

The myriads of Chronic Invalids on every hand, though most of them many times "cured," yet remain a silent but bitter confession of the inadequacy of the ordinary means of preventing and healing disease. The very general prevalence of feeble health, instead of exciting, seems rather to preclude the needed inquiry as to the fundamental principles on which its enjoyment rests; and it must be conceded, in reference to this, more than to any other subject, does every generation seem to defer to the opinions and copy the errors of its predecessor.

The *cause* of such incomplete and unsatisfactory results of Medical Systems and Practice I regard as capable of solution, and not less of correction. It consists largely in the fact, that while medical investigation has confined itself almost exclusively to the single purpose of experimenting with drugs, it has either sadly neglected, or foolishly belittled the eminently potent and fitting means included under the head of HYGIENE.

To estimate the force and realize the consequences of the fact announced in this statement, it may be necessary to give a hasty glance at the general purposes of these two orders of remedial means.   The principles of Hygiene are founded on the truths of PHYSIOLOGY and are sustained and perfected by all the valuable discoveries which mark its modern development.   The established formula of medication are but too generally time-honored medical traditions, and as such require constant modifications.   The one class of means is commended by COMMON SENSE; the other represents the equivocal conclusions of a necessarily confined experience.   It is the function of the one, to respect the laws of vitality, and defer to its inherent powers; the other contents itself too much with experimental toying with these laws, as operating under the disturbances of disease.   While the one strikes at the causes of disorder, uproots the darling vice, and corrects those conditions to which the symptoms are due; the other treats pain as the great enemy, and exerts itself to obliterate or suspend the consciousness of it.   Hygiene has been acknowledged as merely a feeble "auxiliary," when it ought to be classed as CHIEF among the means of cure; and the miserable sufferer has been hopelessly plied with the innumerable expedients of medicine, when he really needed only some of the special and effective adaptations of a systematic Hygiene.   Every person who has been long an invalid, feels to the very core of his being, the want of the remedial plan here proposed; and every such one will regard it as covering unoccupied ground, as supplying his most pressing and indispensable needs, and furnishing a most important but hitherto sadly neglected desideratum in Medical practice.

## SUFFICIENCY OF THE ORGANIC OR VITAL POWER.

Whatever be the theory adopted, or the remedies used, the processes by which health is actually regained, are precisely those which originally maintain it, and they are referable, solely, to the vital power.

The health is directly dependent on the primary act of organization, that of constructing or building up, from elemental meterials of the blood, the organs by which function is performed. Now, this primary and necessary act of growth can never proceed without certain Mo-TIONS occurring in the co-related parts which are the instruments of supply and waste. These motions of solids, semi-solids and fluids, at the same time, and by the same act, eliminate the waste and bring the supply. The identical matters that are brought to the growing parts— oxygen and nutriment are discharged along the same channels, (the capillaries) but in different form. It is at the instant of transformation that the power is manifested. All IMPEDIMENTS to the manifestation of power, that is, to health itself, exist at this point, and the state of the health depends on the degree of perfection with which these na-tural and necessary processes, concerned in producing vitalized forms are carried on. It follows that all healing effects must result from perfecting the relation between supply and demand, at the point of vital change, and hence, actually, by regulating the MOTION of the matters participating therein.

Obviously, then, the wise physician, should consult Nature's meth-od of achieving these results. It is presumptuous to do otherwise. The way in which the "aid" he is so fond of offering "Nature" is to be given, is not indistinctly pointed out, and his legitimate dealings with the living body are restricted within narrow limits. If he would help, he must build as Nature builds, by means of good food convey-ed to the needing parts, by putting those parts to natural use; if he would eliminate, it must be as Nature eliminates, by using the pure oxygen of the air. The sustenance and its effect must be properly and equally distributed by the aid of EXERCISE. No physician, be his ed-ucation or judgment what it may, in regard to the selection of methods of interfering with these natural operations, can hence have any other ulterior object, than to secure the domination in every part throughout the system, of the kind and degree of ACTION appropriate to it. In the view of physiology and pathology here advocated, this should be the *direct,* instead of being the *indirect* purpose of the physician's pre-scription.

A few considerations will enable us to see how powerful is the spontaneous tendency of the system toward health. Health, under all circumstances, is *continued* through the conservative power we have been considering. We only question its existence when it is to be *restored,* instead of continued. We forget that continuance and restoration, physiologically speaking, are identical actions. But, in the great majority of persons, the health, in spite of opposing circumstances, is tolerably good, so much so, at least, as to excite no alarm. The tendency to health has not only to oppose ordinary causes productive of disease, but extraordinary ones also. If we were to estimate the influence of bad air, bad food, bad drink; too little, too much and improper exercise of the different bodily faculties, poisons of various kinds, evil passions and so on, our confidence in this tendency would greatly increase; and we should be induced to wonder what would the spontaneous operations of the natural powers of the body *not* do, if we would but remove the impediments to their action.

Evidently then, every organism embodies as an essential element of its nature, a conservative or healing power, which is, under all circumstances, in silent operation, continuing in a more or less perfect manner its vitality; and neutralizing, more or less completely those causes which are operating to imperil its existence. When relieved, then, of the confusion and obscurity into which various medical partizans have been instrumental in plunging this question, the choice of remedies is essentially resolved into determining what *is,* and what *is not* conducive to natural, equable, efficient, vital action. It is such action *only* that is competent to remove from the system, in an effective though silent manner, and in unrecognized and innocuous forms, the materials of disease. To teach what this action is, is the scope and function of physiology. The idea of a peculiar health-giving quality, with which popular opinion invests drugs, is quite an erroneous one; and even professional notions of their antidotal quality as related to disease, are seriously questioned, and thought to lack adequate scientific confirmation. We have already seen that while entertaining this view, the professional mind seems to remain in the dark as to the power which Nature is exerting, in precisely the same direction in which the healthful effects of remedies are expected.

The whole question of disease and remedies, when divested of scientific verbiage and reduced to terms of plain common sense, is resolved into a simple matter of *nutrition*. As perverted nutrition is the essence of all disease, so perfect nutrition is the cause of health; and the proper methods of restoring health are clearly those means that produce the natural interstitial changes of the body, whether chemical, molecular, organic or dynamic,—that transform, without unnecessary friction or waste of power.

The reader will not understand me as here putting forth any new proposition in medicine. I have only condensed into as few words as possible, sentiments which have been expressed with far more point and freedom by some of the best and wisest physicians of this age. Volumes might be quoted, if necessary, in corroboration of these views. Nor, indeed, are these statements necessary to establish the propriety of what follows, since it can well rest on its own merits; but it does seem important to establish in the mind of the invalid the feeling of responsibility, of a capacity for self-help, and of trust in the laws which rule in his physical being, and thus to lead to a condition without which a "cure," popularly so called, is of very small account.

## EXERCISE, IN ITS RELATIONS WITH HEALTH.

Popular opinion has always accorded to exercise due credit in a general way, for its beneficial influence upon health. Experience proves that those heterogeneous bodily actions, incident to the common affairs of active life, whether of labor or recreation, effected by the co-incident or alternating contraction of opposing muscles, secure, in general, the results to be desired. And it seems to be a wise provision of our nature, that so important a matter should not be left for the slow discovery of the reason, but that instinct should prompt to action, as in the young and unreasoning; or that it should occur under the stimulus of necessity, as in the need of food and raiment.

But when we come to recognize *special* bodily needs, we discover the want of special adaptations of exercise to meet them. Our generalizations are useless in special instances. We now see the necessity for analyzing the combined elements of which physiology is composed, for the purpose of discovering how its special wants are best subserved. Science now takes the place of empiricism, and we discover not only that every movement is a more or less profound expression of physiology, but that physiology is materially controlled by movements. When the will commands, it is not only that the muscles obey, but a train of interstitial and chemical actions of great importance is instituted. If we study the direction and extent of this train, we obtain the desired specific results; otherwise, actions perpetually neutralize each other, and no conspicuous or peculiar beneficial effects are produced. Hence, bodily movements become curative or not, in proportion as we know how to take advantage of the lever which a knowledge of them places in our hands.

We might, perhaps, get a clearer view of both the general and the specific value of exercise were we to inquire the effects of its *absence*. Though not the sole cause of the circulation of the blood, yet it is an indispensable condition. The contracting muscles everywhere impel the blood along its course, and without the aid of the conditions established by exercise the motion of the fluids of the body flags and ceases; there is no demand for nutrition, and the interstitial growth of the system stops; the outlets of the body become choked, and waste matters are not disposed of; affinity of the blood for oxygen declines, and the vital fluid remains unpurified; the stomach no longer digests food, because it is not taken away from that organ; the extremities lose their healthful temperature; the brain becomes oppressed and feeble; the nerves recognise the faulty condition of the system, and pain is a natural consequence. In this condition vitality plainly loses some portion of its control over the mass of substance of which the system is composed. It follows that the non-vitalized materials which form a prominent part of this mass, are disposed to obey the laws of crude chemistry; and inflammations, congestions, or other disturbances of various kinds follow or are perpetuated.

The reader will now be able to understand what is the real condition of the system when it is near to an attack of acute disease; as, also, when chronic disease is present. He will also see upon what the removability of disease mainly depends—namely, functional action. The local characteristics, which generally only afford a name for that which exists by virtue of the state of the whole system, are of far less importance than has often been supposed. In fact, the local manifestation, whatever that may be, can scarcely meet with cure while the morbid state of the system at large is continued, for the latter alone renders such local manifestation possible. Since then, *all* functions are intimately associated with movements of the physical organs, it is clear that in them we have the means of controlling disease and restoring health.

Hence it appears, that Exercise is by far the most important of the different branches of Hygiene, because it *controls* the others. It not only determines the amount of food and oxygen required by the system, but is absolutely necessary in the disposition there made of them. A man may be *poisoned* under a selected and even restricted dietary if the food taken be not duly disposed of; and we have all seen what dietetic outrages the hard worker may commit, with apparent impunity, because that which would otherwise prove noxious, is speedily eliminated under the energetic respiration and functional activity secured by exercise.

## THE MOVEMENT-CURE.

But, powerful for the promotion of the above ends as exercise or bodily movements confessedly are, their beneficial effects for the

invalid have generally been practically unavailable. Their office
has been limited to *prophylactic*, or preventive, rather than curative, ends. The reason is obvious. For movements have been often
rendered by the manner of their performance, either *mis-directed* or
*exhaustive efforts*, which are necessarily followed by the *congestion*
of the parts, and a condition approaching, if it be not absolute
disease. And if disease be present, all its symptoms are likely to be
aggravated, perhaps to a dangerous extent. Nothing could more
thoroughly demonstrate the *power* of the means under consideration
than such evil consequences of ill-advised efforts to carry into practice
principles so dimly conceived. And such experience having proved
to the invalid that *evil* as well as *good* results may follow in some
undecided ratio, he is left in a sad perplexity between the unfui
filled promises from different medical sources and his untoward
experiences.

By the *Movement-Cure*, all this seeming confusion and chaos of principles and practice, is reduced to perfect order   The term *Movement-
Cure* indicates an *organized system* of special and local movements,
based on primary laws of the constitution, and having strict reference
to the diseased state.

By a *movement*, is indicated *motion, produced in either the fluid or solid
elements of the body, or both, whether by means of the vital contraction of
muscle, or by external and purely mechanical causes, or by a combination of
these.* The *will* power may, or may not, be made available, according
to the design, though the term does not indicate the fact of either.
But the term *movement* does embrace the idea of strictly physiological
and nutritive changes of matter, caused by *motion*, received by an
organ, region or locality of the body,—it may be at that point to
which the action is obviously applied, or in contiguous or even remote
portions of the body. It will be seen that the word *exercise* is entirely
inadequate to express what is meant by the term *movement*.

## CONDENSED STATEMENT OF THE METHODS AND OBJECTS OF THE MOVEMENT-CURE PROCESSES.

The Movement-Cure processes are carefully prescribed. It is under-stood by this, that a correct estimate is made of the disease, or what-ever impediment to health there may be; and that to meet and remove this condition a certain *kind, number* and *order* of movements are arranged and applied at certain intervals, until a change in the symptoms demand a change in the kind, arrangement and number of the movements applied. It is further held that a different selection and arrangement of movements would prove less beneficial, or perhaps positively injurious. The means of diagnosis are more ample and complete than those employed by physicians practising under other systems, since the operator not only becomes cognizant of the locali-ties of pain, but the amount of functional power is also rendered manifest by the peculiar tests afforded by the treatment. In the operations of the Movement-Cure, the patient is subjected to a literal *handling*, and elements of disease speedily appear, which otherwise might have long remained undetected. Thus the conjectural state-ments of the patient in regard to his condition, based on his sensations, are reduced to their true value.

Besides this, the constant operator, being so much in contact with his patient, and learning to appreciate the value of every change of *eye, countenance, color* and *expression* generally, as well as that of *touch, quality* of resistance, and the re-acting powers in all their varieties, necessarily acquires a remarkable *tact,* surprising to those unfamiliar with the educational means through which it is obtained. This ena-bles him to perceive and apply just those movements that are pecu-liarly fitted to the case, and so the result seldom disappoints the most sanguine anticipations. No doubt unscrupulous charlatans have availed themselves, to some degree, of this peculiar tact and close observation, to deceive and ruin their victims; but this only proves the importance of these advantages when legitimately and scientifically directed.

### PASSIVE MOVEMENTS.

A certain portion of the movements forming a prescription, are *passive*. In general, the weaker the patient the greater the portion employed of this kind of action. A passive movement is where the patient is the recipient of an action originating in an exterior source and communicated to him. The effect of this kind of movement may be describ-d, as causing the weak, relaxed and distended capillary vessels of the part subjected to action, to *contract;* the blood-corpuscles and other impediments that may adhere to the inside walls of the vessels to become *detached,* and consequently to allow the stagnant contents of these vessels to move along their course, and a supply of fresh blood to flow in. Hence it will be seen, that these actions *remove congestion*, and are equally applicable to any part of the body, whether internal or external. The immediate apparent effect is to remove pain, soreness, fullness, redness and heat, and to equalize and soothe an irritable part. While *active* movements are applied at a distance from the severest manifestation of disease, *passive* ones are, in many cases, applied directly, but with the utmost gentleness, to such parts. The mode and amount of these applications are, however, governed by circumstances. The methods of applying passive movements embrace every conceivable form of manual dexterity. The operator himself, in this case, makes the exertion, and whatever effect is produced, is at his expense; while the patient is strongly impressed with a grateful consciousness of *support*, or *help rendered* and *received,* which seems to be diffused throughout his nervous system, affording hope and courage to the mind, and new vigor to all the organic manifestations. A noted empiric, having adopted the use of a few passive movements, such as *clapping, percussing, shaking,* etc., has lately won a reputation, and, it is said, a fortune. This is due to their effect of immediately causing real pain to subside; and, when skillfully applied, in alleviating soreness, swelling, and other, even severe, evidences of disease.

### ACTIVE MOVEMENTS

Are those in which the patient is required to *do* something. Their peculiar curative effects are obtained by means of the qualities conferred upon the movement by *localization;* by the *resistance* or obstacles to the effort or action : and by the *time* in which it is performed.

By LOCALIZATION, is meant the confinement of the action to a particular region or perhaps organ of the body, while the remainder :s at rest. By this means, the acting part is made to receive whatever nervous and nutritive supply the action demands. The action itself tends to *focalize*, at a particular point, the available *forces* and *materials* of the system. It is a means of powerful control over the circulation of the blood, sending it, in increased amount toward the indicated point. Either of two important objects may be gained by the same action ; it may supply the requirement of the part for more nutrition ; or contiguous organs, suffering from congestion, may be relieved by sending the surplus away. Movements are generally so contrived as to secure *both* of these purposes.

By means of RESISTANCE to the action of a muscle, the effect of localization is greatly heightened; it causes a great (if necessary, the utmost) energy of the will to be thrown upon a particular point in the organism. and so all the nutritive changes at that point are necessarily greatly increased.

In duplicated movements,* this resistance is supplied by the hand of the operator; and, as the effort or action is caused by and depends on the resistance, so it also appears that the *quality* of the power thus evoked depends greatly upon the quality of the resistance afforded. This is radically different from overcoming an inanimate obstacle. It is a *wrestle* instead of a dead lift. There is vital encouragement instead of

---

* DUPLICATED MOVEMENTS, are such as require one or more assistants or operators for their performance. Single movements are performed by the patient himself. The former are almost unlimited in scope, variety and adaptation ; the latter are very limited in all these particulars, and are useful only for persons whose health is not seriously impaired.

a depressing sense of inefficiency. The mode of resistance incites the production of adequate power, for it begins feebly, so as to be *en rapport* with the patient, but gradually increases as the movement advances, until it approximates, but does not quite equal the power of the acting muscle, and then gradually declines until it ceases. The patient is gratified and encouraged to find, that a *well-managed resistance* evokes action at points where it was otherwise impossible, and that his *capacity* therefor, surely and constantly increases. The powers of the system are *aroused* and *directed*, but never exhausted. The patient is never allowed to feel that he has not adequate resources, but on the contrary is constantly made to be surprised at the gradual expansion of his powers. But much of the success of the treatment depends on the *tact* previously spoken of. Few operators acquire it; many seem unable to obtain it. The peculiar modes of reaching and *touching* the interior animating principle of the invalid, hardly admit of description. But the reader may conceive that certain distinct impressions may be made, securing corresponding responses, by various degrees and kinds of resistance—rapid, slow or irregular, at the beginning, middle or ending, of the movement—by tremulousness, or any of the peculiar qualities by which musical sounds are designated. The human system may indeed be compared to a musical instrument, which readily yields harmonious sounds to the magic touch of the skillful operator, but which loses its qualities when imperfectly handled.

The results of the movements are *recorded* in the primary organizing acts, and become apparent in the manifestations of force. These results are at first feeble, but are increased by the every-day's operations, giving the invalid, day by day, a greater and higher control over himself, till it mounts to the highest health of which he is capable.

The TIME, or RHYTHM, with which a movement is performed is, if *active*, slower than ordinary exercise; generally, much slower, but this is somewhat modified by the age of the patient : thus the young permit quicker movements than are allowable for persons more advanced in years. The reasons for slowness of movements are these. In a common exercise, and in gymnastic exercise, but comparatively *few* of the *fibrils* or ultimate elements of the muscles, participate in the contraction. If the time is prolonged, *all* the elements engage; that is, the whole bulk and substance of the muscle contracts, and the advantage is proportionately greater. Experience also proves that the nervous resources are husbanded in proportion as the effort is moderate, though it be greatly prolonged. It is evident that in common exercise the whole system would be rapidly exhausted if all the fibrils of muscles were engaged in action; but when, as in the operation of the Movement-Cure, the action is local, the system can well afford its contributions to the local part, without danger to its resources, and consequently without fatigue, even though *all* the local elements engage. It may also be mentioned, that *time* is requisite for the system to summon its energy and transmit it to the acting point. Besides, the distinction of *quality* in the resistance becomes more appreciable in proportion as the time for its development is prolonged.

Perhaps the most important circumstance determining the effect of a movement, is the *position* in which it is applied. A prescribed position is, in fact, a movement, since it necessitates the continued action of certain muscles to sustain it; putting, therefore, the whole body into positive and *definite* relations with the acting part. But, besides the effects of contraction of muscle required by position, there is another circumstance, not present in ordinary movements, which may be made available as a most important auxiliary to the effects heretofore explained. Reference is now made to the effect of gravitation of the contents of the blood-vessels upon the vessels themselves, upon the organs to which they are distributed, and upon the system at large. When this pressure is increased by the position of a member of the body, the walls of the vessels contract against the pressure, and all its elements take on renewed activity. This is another instance of *resistance* provoking *action*. By taking advantage of the almost infi-nite variety of postures of which the body is susceptible, a corresponding variety of effect is produced. Thus, a person may take either a *standing*, *sitting*, *lying* or *hanging* position, and in each instance not only the gravitation of the blood in the vessels, but the action of the muscles caused by sustaining the weight of the body are changed. But the effect of each of the above positions may be further modified by causing the trunk to *fall*, *curve* and *twist*, in any possible direction and degree. Still further, the arms and legs, either or all, may be brought into any position which the conformation of the joints will allow, limited only by the limits of muscular action. This would still further control the nutritive operations of the body. These effects are secured by position alone. Now, a movement always implies a *position*, as above described, and in addition to that, the *action* of some localized muscle, or combination of muscles of a part already either expanded or contracted by the necessities of the position. It is plain that this combination of causes, each acting as auxiliary to the other, secures in a high degree, the greatest possible effect derivable from combined physical and physiological causes.

It becomes manifest from the above detail, that *movements* produce their effects by simply *supporting the vital motions of the body, at any point, or all points, wherever these suffer.* They convert *pathological* action into *physiological* operation. They aid in the preparation and disposition of the plastic elements of the body, anterior to their assuming vitality. *while* thus endowed, and *after* such matter has lost its vital endowment. It will also be perceived that the effect is to destroy the proximate elements of disease, in preference to subduing its symptoms; while the latter effect is often the ultimate reach of the common methods of practice.

## GENERAL EFFECTS OF MOVEMENTS.

Invalids proposing to employ the Movement-Cure, will be interested to learn what will be their probable experience, and what results they may expect. In the first place, they should be advised that its application is not essentially a *muscle-making* process, but a *co-ordinating* and *harmonizing* one. The new actions that are superinduced imply a new disposition made of nutrition and of the nervous force. There are consequently, *direct* effects, and *remote* ones, dependent on these actions.

The *direct* effects of a single application of a prescription, are noteworthy. There comes a pleasant glow of warmth over the parts most operated on, which generally extends over the whole body; the hands and feet become warm; whatever pain may have existed disappears, or is, at least, quieted; the *pulse falls*, sometimes as much as ten, fifteen, or even more beats per minute; the respiration becomes calmer and more profound; and there is a great disposition to quietude and rest, which should be encouraged. The patient throws himself upon a couch and generally goes to sleep, if he allows himself to do so, and rises refreshed, and with a feeling of renewed vigor.

As the treatment progresses, the sensations experienced will vary with temperament, nature of disease, kind of prescription, and its usual occasional modifications. In some, there is an apparent aggravation of the affection. These symptoms are temporary at most, but generally require the prescription to be modified. Sometimes there is no seeming improvement for five or six weeks, when the patient becomes suddenly conscious of a change in the pulse, skin, increase of the circumference of chest, of the motions of its walls, etc., the assurances of returning health.

A certain portion of patients, in from one to three weeks from beginning regular daily treatment, become affected with *headache, furred tongue*, and *feverishness*. These symptoms are not generally very severe, though occasionally they confine the patient for two or three days to the house. Even vomiting and diarrhœa may occur at these times, without other apparent cause than the disturbance produced by the treatment. Bronchial catarrh is an occasional symptom, and so are eruptions of the skin. In one instance, the treatment was suspended altogether. in consequence of an intolerable eruption which made its appearance, whenever the treatment was resumed. These critical symptoms may be avoided by sufficient moderation; but, to some, they afford encouragement in the conviction which such patients might otherwise be disposed to resist, that a beneficial change is occurring in the physiological system.

The most rational *cause* for the occurrence of these symptoms, appears to me to be the greatly increased action of the *venous absorption*, which is induced throughout the system by the movements. It has been seen that most of the operations contribute to remove the obstructions of the capillaries and increase the flow of fluid therein—the very condition which induces an inward current from without their walls. The fact of the production of muscular and nervous waste, implies this; but the energetic manner in which it is now accomplished, carries along a large amount of non-vital material which was previously lodged in the system, pervading the inter-muscular juices. This being suddenly called into the blood, nearly poisons it for the time, and the vital powers are thereby called upon to assume renewed eliminatory action, thus occasioning the disturbance.

It is to be understood that these *crises* are not desired, and that the gradual and natural resumption of the due relations of elimination and supply should occur unobservedly as they do in perfect health. They do not occur in the great majority of well treated cases; but they have their use, chiefly in a moral point of view, being a slight deference to the long-cherished popular notion in regard to the importance of such symptoms, as evinced by the effort to secure them in some form by the operation of drugs.

## CHANGE IN SHAPE OF BONES.

These changes of substance in the body are by no means confined to the soft parts, but extend to the *bones* as well. The change in the shape of the *spinal column*, the *deformed chest*, etc., will be treated specifically further on; but I will introduce here the following evidence of the changability of the bones.

The late Prof. Rezius, of Stockholm, Sweden, the celebrated Ethnologist, in showing me his extensive osteological collection, pointed out several interesting abnormalities of bony structure, in connection with the history of the persons to whom they belonged. One specimen was from a beggar, who feigned lameness in the right leg from his boyhood. A slight rheumatism afforded the original suggestion, and he spent the rest of his life in receiving alms at the end of a bridge in the city. The result was, that not only the leg but the pelvic bones of that side were of diminished size; the thigh bone measuring an inch in circumference, and the leg an inch and half in length *less* than the unfavored one. Another was from a prisoner, confined for some capital offence, by one leg heavily ironed, who after five years escaped, and was found frozen. The bone of the unused limb was undiminished in size, but as light, apparently, as a piece of pine wood of the same size. A curious skeleton was that of an old lady, whose sole occupation during the last years of life, being in the alms-house, was that of constant knitting. The shape of the bones corresponded with the prolonged position. The back was bent to an oval; the sacrum was horizontal, and its extremity was but ten inches from the top of the sternum, and the transverse diameter of the chest, at the lower portion where the elbows naturally pressed, was but *five* inches, the anterio-posterior diameter being nearly twice as great. The vertebræ of a carpenter were not only harder, but heavier, while those of a tailor were more spongy and lighter than the average; the one carries burthens while the other rests upon his elbows. Numerous — other specimens illustrated to the fullest extent the beauties of tight lacing, and the conformation of the consumptive—the chest-circumference of these latter being, in several instances which I took pains to measure, two or three inches less than others of corresponding height and size.

The habitual movement of breathing, where this is faulty, is also changed by movements. In most invalids, the breathing motions are nearly limited to the superior portion of the body. It is almost always a distinct purpose of treatment, to regain the natural oscillatory motion of the stomach and bowels, so indispensable to their health.

The temperature, and especially the sensation imparted by the skin on being touched is very much changed. This organ becomes cool, velvety and elastic, instead of maintaining the previous hot or cold, but dry and hard condition of surface. It receives more blood, *breathes* more, and is more vital. The treatment, or rather some of the movements which send a rush of blood to the skin, has been called by ladies a *beautifying process,* suggested by the fact above specified. There is confessedly nothing so improving to the personal appearance, as a restoration to health. The carriage is more easy and noble, and more consideration is involuntarily paid to persons in the full possession of all their powers.

From what has been said of the effects of movements, it is evident that *they must be highly curative,* because they bring into requisition the whole of the healing power of the body, upon which alone the cure really depends. It is also evident that *movements furnish a remedy, applicable in every stage of disease.* None are so weak but that some of the infinite modifications of these applications are available. Confinement to the bed or to the recumbent position is no obstacle, neither is want of power for voluntary exertion, since the treatment is kindly adapted to all such cases.

## THE ROOM, APPARATUS, AND MODE OF OPERATION.

The ROOM for the application of treatment should be on the first floor, for the better accommodation of persons too feeble to mount stairs. It should be of good size, to allow of freedom in walking about, with high ceiling, and windows opening at opposite sides, to insure abundance of air. It is necessary that a dressing-room be attached.

The APPARATUS, or FURNITURE of the original Swedish Institute, is simple in the extreme, but as the treatment has become extended to other lands, the conveniences for applying it have increased. As essential articles, the following may be enumerated: Several couches, about fifteen inches high, covered with rept or enamelled cloth, and well stuffed with hair, so constructed that one end may be let down, and the other elevated to serve as the back of a reclining-chair, and instantly changable, so that a person may *sit, recline* or *lie* upon them. Cushioned benches, elevated about three and half feet, narrow enough to sit astride, with places for fixing the feet adapted to the height of different persons, and to lie upon; posts, with pins passing through, projecting on either side for handles at different heights, from floor to ceiling; padded horizontal bars, to support the body at different heights; round bars for grasping, which may be placed at different heights, horizontally; also, perpendicular poles, at various distances apart; instruments for *bending* the body at any point; for pressure upon any desired region of the body; for raising either shoulder; for expanding either side of the chest, and to fix the body at any point, while a passive movement is applied; an instrument for clasping and fixing any portion of the body, while in the standing posture or in the sitting posture; one for applying local passive action to any of the members. Also, apparatus for slow bend-and-twisting the body in all positions, for spinal curvature. The above apparatus is necessary for the purpose of maintaining the desired position while the movement is applied to the prescribed region in the prescribed manner, by the operator.

DRESS.—In all cases ladies relieve themselves of hoop-skirts, and of all pressure of clothing upon the waist or other portions of the body. Some present themselves in loose wrappers, while the greater number supply themselves with short dresses with pants, generally made of plain colored flannel, neatly trimmed with braid, or such plain trimming as fancy may dictate. Happily, in the case of gentlemen, the common mode of dress is the most suitable.

METHOD OF APPLICATION.—After the diagnosis is settled, each patient receives a prescription, indicating what movements he is to receive, and how they are to be applied. To obviate a troublesome amount of writing, as well as to bring the whole direction so immediately under the eye that it can be instantly read, the prescription is written with abbreviations and signs. The patient keeps this in sight till he gets through it. When ready for a movement, the patient selects the assistant he wishes to apply it, who is expected to observe the precise order of movements indicated, and to apply no other, without permission from the physician in charge. The movement is slowly applied by the aid of one, two or more assistants, as there may be need. During the operation, the patient is requested not to engage in unnecessary conversation, for in case of weakly people the movement is sure to falter, and perhaps *stop*, should the patient even speak. This shows that the performance of any act is at the expense of the general system; and also indicates the conditions necessary to be observed in order to obtain a health-giving command of the system, or of its parts. The movement is gently repeated three or four times—more if there has been an inaccuracy on the part of the patient. If it be passive, or a quick movement, there is an indefinite number of repetitions. The efficacy of an active movement seems to be impaired or lost by repeating it many times, because then it begins to induce exhaustion instead of causing an accumulation of vigor. After the movement, the patient, if feeble, sits or reclines; if able, he walks about, or engages in conversation with companions.

The lapse of several minutes is required for the train of actions induced by a movement to become completed. If the next movement is applied too soon, this train is diverted into other channels, and the special effect is diminished; and if the patient or operator, or both, are inattentive, the movement degenerates into a mere common exercise. If, in addition to hurry and carelessness, the movements are badly selected, and the special conditions of the patient imperfectly appreciated by the physician, the patient might nearly as well amuse himself some other way.

Immediately after retiring from the Movement-room, the patient is expected to lie upon a couch for half an hour or more, till the changes induced by the operations have been fully completed. During this period sleep is almost irresistible, and the patient is encouraged to indulge in it.

Movements are generally applied to parties of several persons at the same time, each of them taking his turn, but without particular regard to regularity. They are applied once a day, but those who are at hand may have a few supplementary ones, generally at night. If a patient be too weak to walk, he is carried to the Movement-room, in preference to receiving treatment in his own room; but may receive treatment in his room, if his disease precludes his removal from it.

The social element which reigns in the movement-room is a highly valuable adjuvant of the treatment. This I have remarked in the foreign institutions, as well as in my own. All are anxious to represent to others, and especially to new comers, their evidences of improvement, and to strengthen each other's hope; and an almost irresistible influence for good is thus exerted on the desponding. This is in striking contrast with the discontented and complaining tone of feeling that is apt to prevail in those hygienic establishments where movements are either not used, or so used as to degenerate into mere gymnastics.

PRESCRIBING.—The physician practicing the Movement-Cure, has constant exercise on the living subject, in anatomy and physiology. By practical observation he acquires great familiarity with the system, and he comes to *see*, mentally, the natural working of every structure. He also instinctively connects with certain symptoms and appearances corresponding kinds and degrees of *changed* action. He goes still further in his study of the living body, and endeavors to extend, gradually, gently, but positively, the actions of physiology, through all the appropriate channels, not only to the very boundaries of, but even within the affected or diseased region. He finds by experiment, that this object is best accomplished in different classes of cases, and in different individuals, by very different modes of operating; and he ultimately forms a judgment, and instinctively adopts rules applicable to these different classes; and even applies distinctive methods to different individuals. Thus, in one class, containing perhaps a majority, the return of vigor is best evoked by addressing first the respiratory function, and by peculiar management increasing its action, effectively and safely. Another class is best treated by impressing first the lower extremities and gradually gaining the trunk. Another, will require the powers of the body to be worked outward by addressing first the proximate and afterward the remote parts of all the extremities. For some, it is important to counteract apparent congestion of the spinal cord; and for yet others, a few gentle touches, or strokes along some prominent nerve connected with the diseased region, exercises an almost magical influence upon the feelings, and through them, upon the actual pathological condition.

It is also specially necessary that the *feelings* of the patient, not only be not wounded, but that they be pleased and gratified. It is not only possible, but desirable and important, to turn the *emotions* of the invalid, as well as his will, to the account of his health. And when this is successful, as in most cases it is, this class of mental powers exercises a more powerful and certainly higher influence in the control and correction of disease, than all other influences combined. There can scarcely be a doubt but that the contact of the patient with the energetic and controlling operator, and their tacit unity of effort for the execution of a single and highly desirable purpose, are powerful conditions, tending to successful results.

## AUXILIARY HYGIENE.

As the Movement-Cure consists essentially in an effective adaptation of what is universally recognized as a particular branch of Hygiene, it follows that all the other branches are its natural allies and supports. Those who would receive the most effective and satisfactory treatment, should not think of dispensing with special regulations in regard to this matter. Indeed, the very fact of illness may, in the great majority of cases, be considered as a clear confession of having entertained faulty practices in regard to general hygienic habits. Perfection in one branch should by no means form an excuse for neglect of others. Even though the undeniable fact be insisted on, that this branch does the work of the others, yet its curative efficacy must be, to some extent, compromised by forcing it to assume such a disadvantage. Thus movements, by sending the blood to the surface and extremities, contracting the tissues, opening the pores, etc., secures the object of *bathing*, yet it is undeniably better to solicit the aid of the latter resource, to the extent of its legitimate functions. As respects *diet*, the treatment often causes a spontaneous change in the notions and habits of the invalid. It purifies the secretions, and removes the congestion and nervous irritability of the stomach, as hereafter explained, and therefore corrects the morbid appetite, and restores a healthy relation between *alimentation* and assimilation. The curious phenomenon is of not unusual occurrence, of *decrease* of what was supposed to be appetite with the *increase* of health. Invalids who have considerable flesh are apt to meet with this experience. It is probably explained by the fact of large returns to the circulation of useless matter deposited in the system, of no vital account, to be again subjected to the vital crucible. Lean, dyspeptic people, on the contrary, sometimes find themselves able to partake of food in quantities which would previously have alarmed them. This is obviously in consequence of increase of assimilation throughout the system, induced by the treatment.

These facts should be regarded as showing the wonderful superiority and power of the Movement-Cure, and not as affording any excuse for improper dietetic or other hygienic habits. Indeed, most invalids

that are worth the effort for restoration, will willingly assent to the
importance of availing themselves of *all* means that are truly hygienic,
and will never be convinced of any danger of rendering the preponderance of healthful influences over disease, too great.

## COMPARISON OF MOVEMENTS WITH OTHER EXERCISES, SUCH AS GYMNASTICS, CALISTHENICS, ETC.

For the benefit of such as do not readily apprehend the distinction
between Movements, and other modes of employing exercise, the
following brief comparison is made, at the risk of perhaps repeating
what may be elsewhere in substance, stated.

*Movements* are designed for the *sick*,—*Exercise* for the *well*. *Movements* are *prescribed* in accordance with Pathological indications,—
*Exercises* are governed by caprice or chance. *Movements* are the physician's directions,—*Exercises*, the druggist's shop. *Movements* are hom-
ogeneous,—*Exercises*, heterogeneous. *Movements* are so collated as to
support each other in the production of designed, special effects,—
*Exercises* are desultory, and are incompatible with such ends. *Movements* have a *prescribed* manner, and break up old habits,—*Exercises*
have a *habitual* manner. *Movements* are confined to *designated* parts of
the body, or are *localized*,—*Exercises* allow *all* portions to engage.
*Movements* require intervals to husband their effects,—*Exercises*, neg-
lecting this, prodigally waste the power. *Movements lessen* the frequency of the pulse,—*Exercises increase* the frequency of the pulse.
*Movements* are careful to secure, for the time, a preponderance of
*assimilation* in the system,—*Exercises* are liable to cause unnecessary
disintegration or destruction of organized matter. *Movements* increase
the *arteriality* of the blood,—*Exercises*, if violent, increase the *venousity*
of the blood. *Movements*, for the feeble, are performed by the aid of
of an auxiliary power,—*Exercises* are the visible results of exertion.
*Movements accumulate* nervous force,—*Exercises*, in general, are liable to
*exhaust* nervous force.

# THE PRINCIPLES

OF

# THE MOVEMENT-CURE,

### ILLUSTRATED BY REFERENCE TO CASES.

It is no part of the present design to attempt to win the reader's favor by the very common plan of bringing a formidable array of cases presumed to be evidences of the merit of the treatment. It is the object, rather, to represent to the intelligent reader, so far as the present prescribed limits will allow, the entire *reasonableness* of the peculiar method of medical treatment here advocated; that it is founded in the first principles of physiological science, challenging the criticism and the support of all medical men, of whatever school; that it is entirely practicable; that while it is eminently efficacious in all common cases of chronic disease, it is also effectual in many cases which are otherwise confessedly beyond the reach of medical means. It is quite wrong to infer, as many have done, that the Movement-Cure is useful only for "hard cases." The fact which is now pretty extensively conceded, that it is of *any* value in these, ought to prove to the reasonable mind, that it is of still greater value in ordinary cases. The object, then, in citing particular cases of success, is to bring the *principles* involved, and the necessary consequences of these operations more directly to the attention of the inquirer and not to imply the promise that every similar case, without reference to constitution, habits of body or mind, or moral deserts, will infallibly meet with equal success.

The reader will understand that any attempt at description of the methods of treatment would quite transcend the limits of the present work; if, indeed, that were possible by writing. A few *cuts* are introduced, as the readiest way of affording general suggestions in regard to the processes.

## DISEASES OF DIGESTION.

The dyspeptic invalid refers his disorders chiefly to his stomach, and is apt to confine his curative applications mainly to that organ. But, if we consider the actual state of the system at large, we shall find that the stomach disorder is but *one* among many evidences of ill-health; and we are often at a loss to determine whether the indigestion is caused by, or is the cause of the general ill-health under which the system suffers. However this may be, the futility of treating the stomach alone, by plying it with special applications of any kind, is rendered sufficiently obvious by experience. To illustrate the pathology of this disease, and the *rationale* of the "Movement-Cure," when applied as the remedy, we may refer to an actual case, having all the usual symptoms in a high degree of intensity.

CASE I.—Mrs. T., a young married lady of this city, was recommended to take the "movement" treatment by her old physician, a man of eminence in his profession. Her mother had died of consumption in her early infancy. She was thin and slightly built, with very narrow chest; of a highly emotional temperament, and very feeble muscular development, though she was inclined to use her thin muscles to excess, as is often the case with persons of her temperament. She had been treated through several years of great suffering, with no other than palliative results, for symptoms like these: Excessive sensitiveness of the stomach, generally great pain after eating, palliated habitually by wine or bitters; a constant sense of hunger; by turns, violent palpitation of the heart, her really most annoying and distressing symptom.

As is usual in these cases, the hands and feet were habitually cold; the skin bloodless; the complexion dusky and wan; the respiratory motions, confined chiefly to the upper portions of the chest, were necessarily ineffective. She was incapable of *continued* exertion of any kind.

The reflective reader will observe that here are two classes of symptoms exactly complemental to each other. Thus, the paucity of blood in the skin and the extremities is connected with congestion of the internal organs, especially the stomach and its appendages; shrunken capillaries of the external parts, with dilated vessels of internal parts; cold hands and feet, with insufficient respiration; excess of nerve-action, sensitiveness and pain, with defective muscular action; morbid quality of the secretions, with insufficient reduction by oxygen of the waste of the system, consequent on the restricted motions of the chest-walls; prolonged retention of aliment in the stomach, with insufficient action and demand in the exterior parts; morbid appetite, with residual food in the stomach, causing local irritation, and the morbid and painful manifestation of the emotional nature, with the partial closure of the common channels of force.

What has the administration of any drug whatever to do with rectifying such a condition? Plainly, the stomach, with the morbid secretions, residual food, excessive development of nerve power, &c., is already overburthened and abused; and the dictate of reason and science would seem to be, to *relieve* it of the severe embarrassments under which it evidently labors. For this purpose movements are admirably fitted. Both active and passive movements were applied to the extremities; movements were cautiously given to act upon the liver, and to promote the respiratory function; and they were made to succeed each other in proper order, number and combination. At first only temporary and undecided relief was experienced. But these, at first moderate effects, are sure to accumulate by the repeated and regular applications, till they become conspicuous and convincing. There is seldom any lack of perseverance after the first two or three weeks.

**Fig. 1** — BENDING AND STRETCH-
ING THE FOOT—increasing the heat,
and the amount of blood in the extremities, and lessening it in the upper portion of the body.

In this case nearly six weeks had elapsed before the evidences of improvement had become indubitable. At this time the chest had increased one inch in its cir-cumference, and the hands and feet had become habitually warm, and the pain-ful sensations of the stomach began to meet with relief. More color came to the skin, and the eye began to sparkle, indicative of returning vigor and health. She gave up drugs from the first, which served to increase the nervous irrita-bility, but the accumulating evidences of increase of power gradually overcame this symptom. The spells of palpitation* became less noticeable, and then ceased; and in about three and a half months she stopped coming to the institute, having renewed health, and the assur ance of possessing the substantial basis for its continuance. The measure of the chest had increased *two and a half* inches.

During the first portion of the treatment patients are requested to partake of only a moderate amount of plain food, using but little meat, at most but once a day, and to eschew all sweets and artificial beverages.

---

* I was informed by Prof. Satherburg, (of Stockholm, Sweden) that a large number of those presenting themselves at his institute for treatment, had had their complaints diagnosed HEART-DISEASE, and that nearly all recovered. If these symptoms, so alarm-ing to the patient, depend on the irregular circulation, excitable nerves, and the flatu-lent stomach of dyspepsia, there is excellent reason why they should disappear by a treatment so eminently adapted to this complaint. But if heart-disease really existed, Prof. S. rightly contended there can be no more judicious or appropriate treatment for its remedy than this, which carefully but surely removes the excess of blood from the weakened organ to the extremities. and thus allows the heart the best conditions for recovery.

In the different cases of this affection, there are found a great variety of differing symptoms, each calling for modifications of the treatment, to adjust it to the emergency; but the cure in all cases, depends on the same general causes. Thus, in one case, there may be excessive flatulence; in another, heart-burn, evincing spontaneous acid fermentation in the stomach; in a third, eructations; in a fourth, a thickly furred tongue; a fifth will have bad taste in the mouth; a sixth, soreness at the pit of the stomach, not bearing pres-

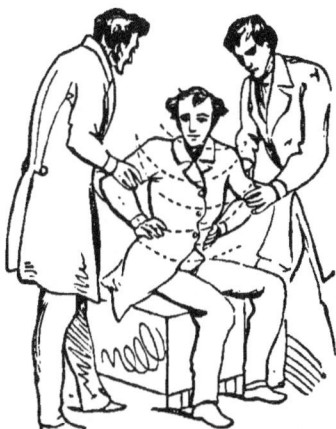

Fig. 2. — Twist-shaking of the Trunk—passive—Conveying to the surface a large amount of blood, and increasing the temperature and function of the skin.

sure; a seventh, will have ringing or other annoying sounds in the ears, and so on, almost without limit; while a majority complain of several of these, besides loss of strength. Now, it is evident to the reflecting mind, that the same cause, and nearly the same absolute condition, which has been to some extent depicted above, exists in every case. But if the system be made to receive a single ounce of oxygen daily more than its wont, it is plain that all the symptoms would be mitigated; the secretions would be partially or completely purified, because the *dead substance*, which in all cases of disease pervades the system in excess of its toleration, would be *destroyed*— reduced to innoxious carbonic acid, water and urea, and find, in this state, certain and easy egress from the system. So, too, if the excess of blood held by the relaxed and distended vessels of the interior of the body, be pressed forward to the membranes, glands and skin, which so much require it, a radical change in the direction of health is at once effected. Similar reasoning is applicable to the distribution of nervous force, and, indeed, *all* of the physiological operations.

Fig. 3.—ARM BACKWARD PRESS-ING—ACTIVE—Turning out the ribs; strengthening and expanding each side of the chest in succession.

The reader will perceive that the symptoms connected with dyspepsia, are more or less associated with diseases bearing other names. This indicates a similarity of origin, and that similar means are required for effective removal. There being a fault in nutrition and assimilation in all, shows that those hygienic conditions which best control these, are required in all. In every case, the *kind* and *quality* of the applications are dexterously adapted to the strength as well as the disease of the patient, and are changed as these change. This being understood, we may be allowed to omit, without much detriment to the reader's understanding, the something of general detail, which has been entered upon in the present instance.

## CONSTIPATION.

CASE II.—Mr. S., a retired merchant, over fifty years old, had been troubled with severe constipation for over a dozen years. During the first portion of this period, he was in the habitual use of aperient drugs, but the symptoms rather increasing, he became convinced not only of their uselessness but their prejudicial effects, and resorted to enemas as a substitute. To this means of daily relief he was a constant slave, for scarcely once a year did his bowels act effectually, except under compulsion. Later, other symptoms appeared, more threatening in their aspect, especially a troublesome pain and feeling of weight in the region of the heart, and he became seriously apprehensive of the consequences. All this induced a desponding state of mind, conflicting strongly with the success of the efforts toward restoration. He became sallow, shrunken, flabby in muscle, nervous and very weak, especially about the loins and lower portion of the body.

Here the general indications of treatment correspond very closely with those of Case No. 1. But there is, in addition, a special symptom, arising from inactivity and innutrition of both muscles and nerves of

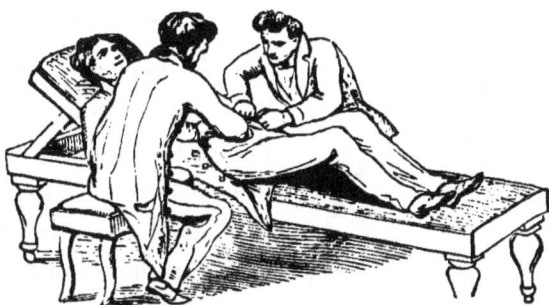

Fig. 5.—A MODE OF ABDOMEN KNEADING—PASSIVE— Increasing nutrition, removing congestion, and exciting the natural peristaltic motion.

the abdomen and alimentary canal. Hence, movements must be so selected that the effects of their action shall converge at the abdominal region, so as to raise the nerves to their normal standard of excitation, and the muscles to their normal vigor of contraction, and that both elements be sustained by due nutrition. In many cases there is, also, undoubted congestion of the mucous membrane, which needs to be removed before the secretions of the alimentary tube can become natural and healthy.

In the accomplishment of these ends, it is absolutely necessary to restore those *natural* movements of the digestive organs, dependent on respiration. The reader will recollect that these organs are located between the diaphragm on the superior side, and the muscles of the abdomen on the inferior, and that every inhalation and exhalation, causes a considerable *oscillatory motion* of the whole mass. In constipation, the extent of this motion is much lessened, generally from prolonged sedentary, and other un-hygienic habits. Hence, it is indispensable to a thorough and complete cure of this complaint, that the muscles concerned should be strengthened, that these natural and involuntary but very essential motions, may be restored.

As concomitant of inefficient respiration, and depending chiefly on it, is a congested state and sluggish action of the *liver*, which nearly always exists in these cases. It devolves on this organ to complete

Fig. 4.—CLAPPING THE
SIDES—Sending a rapid succession of vibratory waves
through the body, affecting
the superior digestive organs
and appendages.

what is ineffectually performed by the lungs—
the purification of the blood. There is generally much feeling of heaviness, and sometimes much tenderness in the vicinity of the
liver, requiring particular attention; though
it is sure to be relieved as soon as the respiration becomes effective.

Movements are eminently adapted to secure the above several objects, completely and readily, and to lead the patient,
long bewildered by the supposed mysteries
of pills and powders,—to wonder how any
other way of restoration ever came to be
thought of.

In the present instance, the results of treatment, though somewhat tardy, were every-
way fortunate. Power was gained in every
part, in proportion as it was directed thither
by the combined efforts of patient and physician. Nutrition, sent by movements to the
regions and among the tissues in which
it was required, fulfilled its duty of strengthening muscle, energizing
nerve, and providing secretion.

As a free circulation of the blood became established in the extremities and among the tissues of the body, the troublesome symptoms
connected with the heart subsided. He thenceforward acquired improved spirits, and more courage and confidence. In about six weeks
his strength had so far improved as to allow his bowels to perform
their natural office, and has since maintained the command of the
function in question, except rarely, when transiently ill from other
causes. He has since reported himself frequently, and continues
well.

The invalid reader will be gratified to learn that the cures obtained in this method are *permanent;* for it is sought in not only securing actionof the bowels, but in removing the *cause* of *inaction.*

## DISORDERED NERVES.

Closely allied with impaired digestion in history and general symptoms, is that feeble state of the body accompanied by distressing *nervous sensitiveness* and *excitability.* This phase of disease does not seem to relate so much to the digestive organs themselves— though these are more or less implicated — as to the *assimilative* function. A wrong disposition is made of nutrition. It seems to support one class of powers, to the serious detriment of all the others. Cases of this kind, passing under various names, are numerous in our existing social system, and very many present themselves for treatment. To illustrate still further the value, and indeed, the entire relevancy as well as efficacy of the Movement-Cure, I cite the following :

CASE III.—Mrs. W., aged about twenty, was induced by a friend, who had also been a patient, to come to me for medical aid. For the previous fifteen mouths she had been confined to the house, and during much of the time, to her bed. She had been in a feeble condition for some time previous. She was brought to the institute on a bed. She had very cold and very moist feet and hands; was very poor in flesh; very weak, with little appetite; constipated ; suffered considerable pain in the back, which presented marks of severe scarification ; excessive nervous excitability, breaking out in tears with the least shadow of a cause; and she had almost entirely lost confidence in her own powers. There were also indications of scrofula.

The cause of this state of things is to be found in part, in too little physical training, in a constitution for which it was very necessary, and in part, in too much intellectual and social cultivation. The necessary results of such faulty nurture are plainly these : there can be no

vigor of digestion and nutrition, because of the slight demand made
in the body for its results ; respiration is feeble, and the chest becomes
narrow, because *exertion* of the physical voluntary powers is always
necessary to give depth and force to this function and suitable
development to the chest. Some form of scrofula is an almost necessary consequence, unless prevented by an excellent constitution
originally. Add to these facts the important circumstance that the
*emotional* nature was under constant exercise, with study, light reading, and exciting social relations, due, not to design, but to our faulty
civilization, and the disorder is rationally and abundantly accounted
for. The little nutrition which the combined physiological operations
could command, was expended in supplying the demand made by
brain and nerve ; the necessities of the organism at large were quite
unsupplied, and the functions. of course, languished. To use a truth-
ful and more direct expression applicable in these cases, the system is
*starved*. And this in a hopeless way, for it is not *more food* that is
directly needed,—food could not be digested. because of the general
lack of organic action and of respiration ; and food in the absence of
active demand for it, is soon changed to poison by its own spontaneous chemistry.

The needs of this case, and of all similar ones, plainly consist in a
radical change in the nutritive habits of the system. The flow of the
blood and of the nervous power should be directed into new channels,
viz : the muscles and other mechanical and physiological constituents
of the body. The chest must be made to heave and food must be
digested and *used*—not through stimulants and tonics. for these at best
only *seem* to effect the desired results ; but by simply *acting*. This is, of
course, done at first by means of efficient external aid. Further, the
*will* of the patient, which has long proved too feeble, needs as much
support as the body. The effort, then, must commence *outside* the
patient, but it must receive little additional rills from the invalid—just
as much as can, for the time, be well afforded, and *no more*. Use,
strengthens ; repetition becomes habit ; and we soon find our power-

less patient using *exertion*, followed by commendable, and what is important, *commended* exhibitions of power.    Immediately the tears dry, courage increases, nervous suffering ceases, and our patient is in the high road, long but wrongly sought, many times but foolishly directed —toward HEALTH.

In a fortnight our patient left the bed; in a month she began to go out with help.   In two and half months she walked regularly, and that without fatigue or complaining.   In four months she returned home to the great joy of her parents ; and we afterwards learned was soon engaged in matrimony.

Fig. 6.—FULLING THE ARMS—PASSIVE— To move the blood in the capillaries , returning it to the heart, and allowing a fresh supply to flow in, and so to increase the general nutritive actions of the feeble parts.

It is unquestionable that in all cases of this kind, the *will* needs skillful and effective training to lead it forcibly in the direction of the muscles, to support the important acts of primary organization.   In this way, the feeble body is made to cure itself; its appropriate actions are increased, and directed ; the system iseffectively enlarged and harmonized.   No officinal preparations, ever so cunningly devised, can pretend to secure the same objects.

Affections of the nerves are so common and so various, as perhaps to justify a still further illustration of their pathology, to render the inference as to the proper remedy more complete.

CASE IV.—Mr. R., of N. J., aged about forty, was conveyed to my Institute in June, 1860.   He was a gentleman of excellent parentage, and of a naturally sound constitution, having had, until now, no sickness from his youth up.   He was of short stature, had a full body with large chest, and was exceedingly well-proportioned.   I found him reduced to an extremity of weakness, with a shrunk, wild countenance, flabby flesh and greatly emaciated; hands and feet cold, wet and inelastic; tongue thickly coated with a white fur; he could take but little food, and he had severe constipation; but he complained of little pain, and that chiefly in the back and the region of the diaphragm; the most serious source of discomfort being the *total suspension of sleep*. According to his own testimony, he has not been conscious of sleep for several weeks, which certainly indicated a formidable derangement of the nervous system. He had, for the last eight months, been under the care of a celebrated hydropathic physician, under which treatment he had acquired the chief characteristics of his disease.

As I have at nearly all times, patients similarly afflicted, who have previously received medical directions from the various *pathics* of the day, with, in the majority of cases, only equivocal and temporary relief, and generally a permanent increase of their afflictions, it is due that an explanation be given of the reasons for such indifferent success.

The distinction between aliment and *drugs*, and their respective offices, seem to be imperfectly understood. Vital structures have an affinity for the *one*; and manifest a repugnance to the *other*. The consequent act in the one case we call *nutrition, assimilation, growth*; in the other, the *operation*. In both cases the living parts and vital powers are called upon to *act;* in the one case to *construct* the organs of the body; in the other, to *repel* the presence of what is not only non-assimilable, but tends to subvert and destroy. The immediate consequence of the peculiar action in any particular case, may be agreeable; but however great the seeming temporary necessity for such action, the effect of its prolonged continuance can be no other than to prevent nutrition, wound the delicate susceptibility of the nerves, and through them, even consciousness itself becomes implicated, and the return to health is rendered constantly more difficult. The increased activity of the nerves induced by drugs, necessarily turns the current of supply to them, on the principle before explained; so that the actual powers of the body sadly fail, while nervous susceptibility is increased to an overwhelming degree—a state which in itself, is a disease of a formidable character.

The Water-Cure treatment, though professing to confine itself to the principles and measures of Hygiene, is yet practically liable to the same objection. The object of the baths is to affect, primarily, the sensory surface. But if these be administered too frequently or injudiciously, or for too long a period, a similar morbid state is induced, and for similar reasons. Every application is an appeal to the spinal centres, and through them to the whole sensory system, rendering nerve-action compulsory. This may not be, and is not always consistent with the vital harmonies. Through these processes, the sensorial order of faculties acquires an undue cultivation, sensorial power is correspondingly developed, while this very action and development serve to restrain organic action in other parts more important in the economy, and even to cripple life itself. The same ultimate consequence ensues as has been previously described.

The Movement treatment is amply competent to respond to every medical requirement of cases of this kind, while it is open to no such objection. In the present instance, the application of passive movements to the extremities, inciting anew the processes that had for some time well nigh failed of action, combined with such mild active ones as effectually directed the nervous currents *from* the oppressed centres, in two or three days caused sleep again to refresh the patient in question; and to his surprise he found himself soon able to walk short distances. Care was taken that no overdoing should bring the least reverse of feeling. The movements, in these cases, are always modified with sufficient frequency to keep pace with the enlarged capacity of the patient. The surface of the patient began to acquire a natural appearance, vivacity to return to the eye, confidence to the mind, and selfhood to the whole man. In four or five weeks he was able to enjoy a fair amount of natural and refreshing sleep, and the white fur of the

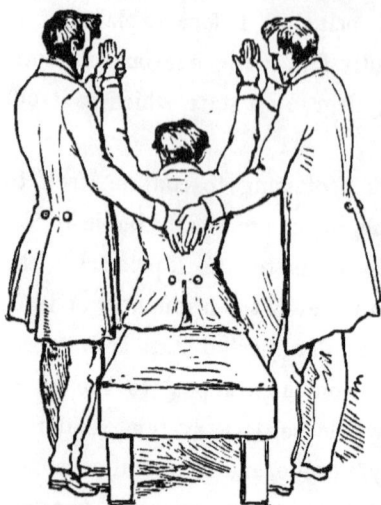

Fig. 7.—TRUNK RAISING, BACKWARD BENDING, and BACK STROKING —ACTIVE AND PASSIVE —Affecting the abdomen, the diaphragm and the spinal nerve.

tongue in good part disappeared. In two months he had quite changed in appearance. As his worldly affairs had been settled preparatory to his expected speedy dissolution, he was at leisure to remain and persevere in the treatment for three or four months longer. At last accounts, he was pursuing his ordinary business, and quite well.

The reader cannot fail to see the importance of the principles illustrated in the present connection. Their timely application would hardly fail to *prevent* the greater part of the chronic disease so rife in our midst; while they at the same time supply a philosophical and more effectual remedy than has heretofore been available.

## CHLOROSIS.

Popular, as well as professional opinion, has accorded to Exercise in general, a good reputation for the cure of this form of disease. This is not because it is more appropriate, but because the want of localized disorder renders skill less necessary for safety and success in its use. A somewhat inadvertent testimony to the superior value of this agency was given in my hearing by Prof. Bigelow, Sr. In proceeding with the class in pursuance of clinical instruction through the wards of the Mass. Gen. Hospital, and noticing the conspicuous evidences of this affection in several cases present, Prof. B. remarked:—That girls were often treated here as long as the government of the institution would permit, and then discharged without benefit; but he was often surprised and pleased to meet the same persons at service in the families he attended, ruddy with health. *Work* had done for them what the most approved administration of drugs had utterly failed to accomplish. The reasons why the treatment by movements is curative, will become apparent by reference to

CASE V.— Miss L., of R. I., aged twenty, had been obliged to leave school, and for several months continued to decline in health and strength. She was very weak, being incapable of walking more than a short distance ; had *constant* and *severe headache ;* lips and cheeks quite without color ; backache ; dyspepsia and constipation ; had not menstruated for several months ; pulse 110, with obscure symptoms of pulmonary affection.

In this case there seemed to be an almost absolute deficiency of the power of assimilating food, and adapting it to the purposes of the system. The menstrual

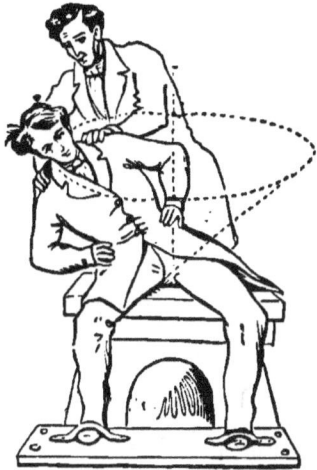

Fig. 8.—ROTATING THE TRUNK IN A SITTING POSTURE — PASSIVE — Strengthening the spine, and producing a gentle increase in the action of the abdominal contents.

function was suspended from sheer poverty. The head ached because unsupported in its nutritive wants. The blood was too poor to attract oxygen ; and in attempting to attain what the system was being ruined for the need of, it pressed with great vigor through the aerating capillaries, consuming, in the effort, the available strength of the system. The blood lacked *iron* as well as other nutritive elements, because the system had no power to extract from the food that which it was designed to supply and contained in great abundance. It is not *materials* that are lacking in these cases, as popularly and professionally contended, but the power of appropriating them. Movements cause the demand to be effective, and are necessarily followed by an increased transmission of nutritive material to the active parts.

In this instance the headache gradually abated, the pulse fell by degrees to the normal standard. In about six weeks menstruation was established, not because it was forced, but because the system had become able to support it. The next period was somewhat delayed, but the treatment was continued, the strength, color, &c., improved, and in about three months the health was quite restored.

Other cases might have been selected for the present illustration, in which the cures were much more rapid; but as this was a most difficult case, belonging in a social position where this form of disease is very common, too often terminating in consumption, it answers well the objects for which such illustrations are adduced.

## UTERINE WEAKNESSES, DISEASES AND DISPLACEMENTS.

Mrs. H., of Mass., was brought to New York on a bed, being unable to sit up, her disease (of fifteen years duration) having recently experienced an aggravation. She had suffered from prolapsus, and the multitudinous weaknesses that are invariably connected therewith. She had for some time been *quite unable to retain the urine*, which consequently dribbled away almost constantly. An examination showed the upper and large end of the womb bearing hard upon the bladder, while the mouth projected backward and upward. She had borne five children, had a pendulous abdomen, and excessively weak and flabby muscles, and for a long time had been unable to get refreshing sleep. She had been treated at various Water-Cures where such complaints are made a speciality in the aggregate about three years, without permanent benefit, and with a similar result as regards the nervous system, to what has been previously described (pp. 35—39).

Here was a case in which the order of development of the complaint was first, general weakness, afterwards, prolapsus—terminating in extreme anteversion. In these different stages of disease, it had defied all medical treatment, and instead of being cured, it had acquired new characteristics, rendering a cure much more difficult.

The Movement-Cure divests complaints of this class of all their ambiguity, and supplies a radical and perfectly successful mode of treatment

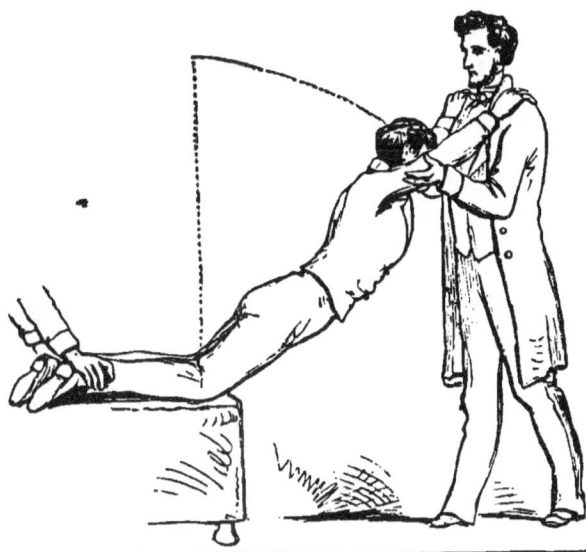

Fig. 9.—TRUNK FORWARD FALLING—ACTIVE—Increasing the space in the superior portion of the abdominal cavity ; decreasing it in the inferior portion, and elevating the contents.

for all kinds and degrees of the various modifications of these affections.

It is superior to the folly of employing artificial supports of any kind, since Nature has provided so much better means; and it also exposes that of *re-placing*, except in peculiar and extreme cases, for the purpose of temporary relief. The womb *floats*, with other visceral parts, in the cavity of the body, and its position depends on circumstances easily controlled. If it be desirable for the organ to rise higher, it is only necessary to *make room* for it, and it will assuredly assume the new position. If the digestive, bears upon the generative intestine, and the whole presses inordinately into the pelvis, and even upon the perineum, it is simply because there is insufficient room above. If now, the interspace between the floating ribs be increased, since the cavity is always full, the pressure is removed from below; and if, in addition, the walls of the lower portion of the abdomen be at the same time contracted, the tendency of the visceral organs to *rise*, is followed up by pressure from below, and a new posi-

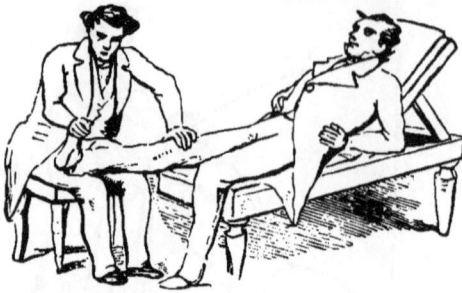

Fig. 10.—THIGH TWISTING—ACTIVE—Removing congestion of the pelvic organs.

tion of the aberrant organ is insured. These sustaining conditions are rendered permanent by the processes of development.

The other faults, those of digestion, of the nerves, &c., were treated on the principles previously indicated.

Mrs. H. began the treatment disheartened and sceptical; she had been furnished with no opportunity to cultivate other sentiments. She soon began to mend. She left the treatment in four months, completely restored, both as regards general bodily health and local symptoms. She desired me to refer to her case, whenever it could be of any service to her sisters in affliction.

The different varieties of affection comprised under this general head, such as retroversion, prolapsus, lateral flexion, ulceration, hyper-trophy, &c., are treated by the almost infinite modifications of which this treatment is susceptible, based on the same general principles.

The following case is introduced to show that the cures thus brought about are permanent, because they imply self-management and self-control; and also, that correct ideas on the part of the invalid are of much greater service than the ordinary supports and drugs combined.

CASE VI.—Mrs. E., a young married woman of a neighboring city, had suffered a miscarriage some three years before I saw her, from which she dates her ill-health. This accident was followed by weakness requiring medical advice. In spite of medical assistance she was obliged to assume the recumbent posture nearly all the time. During the last year she wore a pessary, or internal supporter, this being considered necessary for her comfort and cure. She was removed for change of air to a watering place near New York, when her husband visited me for advice. Being naturally ambitious and hopeful, she experimentally practised my directions, aided by my book. She immediately commenced gaining in strength, and experienced the desired specific effects. In four weeks she became able to walk about a little, and to sit up a good deal. She then came to the Institute, when I saw her for the first time. In a few days I was able to remove the pessary, without subjecting her to inconvenience. She received treatment *two weeks*, and then felt well enough to return to her home, to practice what she had learned. The next time I heard from her she was pursuing her usual employments, enjoying comfortable health.

In a recent letter, Mrs. E. says :—" I have had no physician since I saw you (sixteen months ago), whereas for twenty months previous I was constantly attended, and I am gaining all the time, though slowly. When I over-fatigue myself, or am in any way unwell, I rely upon the movements to restore me."

The Movement-Cure serves to dispel a good deal of the illusion with which this class of infirmities have been regarded. Women have been victimized and the empiric rewarded, with little substantial good to the suffering party, long enough. So far as the influence of this treatment goes, an end will now come to the mistakes under which the invalid's mind has labored. It shows women fully *why* they have suffered from ill-health, and places the preventive, and in many cases, the curative means in their own hands. They may henceforth, in great measure, avoid those indelicate manual attentions of late years regarded in the current practice as indispensable. Many voluntary testimonials, from persons I never before heard of, have reached me regarding the value of the principles and means here set forth, joined with thanks for the complete efficacy of the few suggestions for self-treatment found in my work.*

CASE VII.—The following note, just received, I insert as a specimen of those referred to :

* An Exposition of the Movement-Cure, by the Author.

—— ——, Oct. 16, 1861.

Dr. Taylor :—

. . . . . I know of no way in which I can better acknowledge the great indebtedness I feel to the Movement-Cure, than by allowing you to refer to me and to make my case public, with the hope that by so doing I may induce some feeble and useless woman to seek relief from a disease which I believe drugs cannot cure.

I may truly say I was useless. It was only with the greatest difficulty that I could ascend a flight of stairs ; could not ride or walk without producing congestion of the womb, most distressing. There was great muscular weakness, so that slight exercise would produce the bearing down and sinking at the stomach, so well known to every one suffering from falling of the womb—and violent backache. On rising in the morning I would feel quite comfortable, but walking about the room a short time and making my toilet, would quite exhaust me, and I would have to lie down. Was forced to lie down most of the time, my muscles thereby becoming weaker, and I more discouraged.

At this time your book,* on "The Movement-Cure," was put into my hands. I read it, but thought I could do nothing with it myself, so I laid it aside. A week or two after I had an unusually severe turn, brought on by riding. Was in much pain, with soreness and heat in the pelvic region ; could scarcely walk across the room ; had been growing worse for three days. As a last resort I selected and applied a few of the movements which are designed to remove congestion. The pain left me at once. At the end of an hour I felt relieved in every respect. Hope again sprang up. I continued the movements every day, and at the end of a week the change in my appearance and manner of walking was remarked by my family. After four weeks' application I felt like a new person. Could go up stairs without any difficulty, and could both walk and ride. I have now practiced the movements about eight weeks. The muscles in the region of the abdomen seem hard and are quite able to supply the place of a supporter. . . . . .                    Yours, &c.                    K. N. R.

## PAINFUL MENSTRUATION.

The efficacy of movements in affections of this class, with the reasons therefor, will appear in the following

Case VIII.—Miss C., daughter of a physician in Conn., had been for years habitually confined to her bed for at least a week out of every month, in spite of the resources of professional skill, quickened by paternal solicitude. She was, of course, far from being well the remainder of the time, being nervous, weak and low-spirited, and

---

* Dr. Geo. H. Taylor's Exposition of the Movement-Cure.

having pain in the back and head; but the overshadowing manifestation of ill-health was in the function alluded to.

We may safely assume that *local congestion* was the most probable cause of of the nervous irritability and other attendant effects in this case. The success of treatment in cases of this sort, evidently depends on the same principles as have been previously insisted on, namely, their power to effect the equal distribution of the blood—conveying it from certain suffering portions of the body to other portions, where there is evidently a deficiency. Cold hands and feet, and an inactive skin are always conspicuous concomitants of the form of disease which we are now considering; and in no case can this •bject—the removal of congestion—be more easily and certainly accomplished than in this. The continued existence of local congestion of the parts now considered, must result in those very common, but very difficult forms of disease to cure, *hypertrophy, induraction, ulceration,* &c., which have become the opprobium of the medical art.

The ordinary medical plan in these cases, is that of *local depletion*—effected in one way or another. The *principle* is manifestly wrong; for if the fullness be reduced without strengthening the vital power of the part, it only causes a further determination of the blood to the already overburthened organs, thus *continuing* instead of *curing* the disease. The effect of the movements is exactly the reverse of this; hence their success. They not only convey the blood away from the region suffering, but at the same time afford it the necessary *tone.* In the present case, at the end of a month from beginning treatment, there was evidence

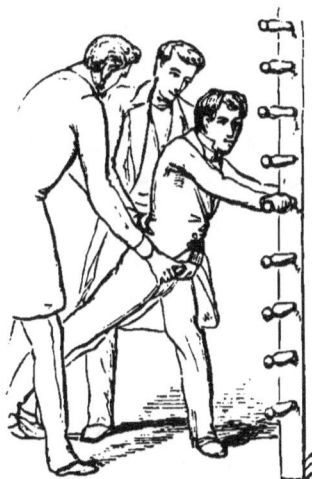

Fig. 11. — TRUNK FORWARD BENDING, and ABDOMEN PRESSING —ACTIVE AND PASSIVE—Strengthening the muscles of the lower part of the trunk, and removing internal congestion.

that a good beginning was made; at the end of the third month she found that her long-continued difficulty had happily vanished. She remains well to the present time.

. In these cases, diffusive stimulants are very generally employed with I am confident, no permanent benefit, but with very prejudicial ultimate consequences. Their influence is as deceitful in this disease, as on any other occasion in which they are employed. If tonics, iron, &c., have a good reputation, it is rather due to the air and exercise which generally accompany the prescription of these drugs.

## CONSUMPTION.

This malady creeps so stealthily upon its victim, that it is often far advanced toward a fatal issue before its nature is suspected. . In the estimation of the invalid, it is, at first, only a little weakness, delicacy, or innutrition, of no account, or not sufficient to induce him to so change his voluntary habits as to secure better nutrition and free scope of his physiological powers. The tendency to unfortunate results is greatly increased by the seductive and deceitful influence of the thousand and one nostrums that are so temptingly brought to his notice, in accordance with the popular delusion in regard to the curative efficacy of drugs in this class of complaints.

A large experience with consumptive invalids, has convinced me, that the scientific application of *Movements* has a much higher claim to be regarded as a *specific*, in every stage of the disease, than any other remedies or combination of remedies, heretofore brought to the public attention. The reasons for the appropriateness of Movements are perhaps more obvious in this, than in other forms of chronic disease. The patient feels that the remedy reaches and corrects the constitutional fault upon which the disease depends. How this is effected, will appear by recurring to an example, such as

CASE IX.—Mr. G., President of a manufacturing company of a neighboring city, had been gradually failing in health for the previous two years, having all the symptoms of phthisis in a marked degree. When examined, he was incapable of longer attending to business, and had the following symptoms: emaciation; pulse over 100; rat-tling sound through a considerable portion of the lungs; inspiration and expiration performed in equal time; feeble appetite; violent cough, especially on rising—it generally requiring an hour to dress himself on account of the cough, which was attended by raising con-siderable yellowish matter.

In these cases, it is not only the cough and expectoration, but the depraved character of the nutritive process—the defective energy of the vital transformations, especially the organizing process—which needs to be removed. This fault is mainly attributable to imperfec-tion and incompleteness of the ærating function. The local malady originates in this circumstance, and depends on it. *There is too little air-space in the chest ; and too little motion of its walls.* Food does not afford strength to the system at large ; it only supports the ineffectual attempt of the system to attain more air by the quickened action of the arterial system. The state of the pulse demonstrates that Nature is striving to bring the blood more rapidly and effectually in contact with the air; under the circumstances the only compensation to be had, but obviously a great waste of force. Instantly, upon the intro-duction of an increased supply of air, the pulse falls; and it becomes permanently lower in proportion as the natural motions of the chest are restored and its measure increased.

Now, there are no means of increasing the size of the chest, the mobility of its walls, or both, so direct and appropriate, as judiciously applied movements. That the desired contact of the air with blood, and through it with the system at large, is rendered more complete and efficient, is easily and convincingly proved. There results a deeper, calmer respiration, affecting equally the whole chest, lessening the fever heat of the skin ; a *fall of the pulse from eight to sixteen or more beats per minute.* These effects begin with the first application

2

Fig. 12.—Arms high ro-
tation and angling—passive
and active—Causing expan-
sion of the upper portion of
the chest, and removing congestion from it.

and increase with every succeeding one. The necessary consequence of this is better elimination, a more perfect nutrition, increase of available power, and decrease of local symptoms. I believe that the formation of tubercles is successfully opposed by this means, and that they may even be caused to disappear. At least, there is in my possession abundance of evidence afforded by a great variety of differing cases, that would seem to unite in proving this proposition. In the case under review, the measure of the chest increased in four weeks from thirty-five to thirty-seven inches; and there was a corresponding increase in the motions of it walls. The consequence of this change was manifest in a commensurate improvement in the health. His treatment began on the 14th of July; late in the fall he resumed his business; eighteen months afterward I heard of his continued good health.

The most gratifying circumstance in relation to this mode of cure, is its *permanence*. The inference which is naturally made from the *nature* of the change effected is corroborated by experience. I may mention the two following instances in proof of this fact. One was a functionary of the Swedish government who, having all the usual symptoms of the disease in its confirmed state including bleeding of the lungs, at the recommendation of his physicians resorted to the Movement-Cure Institutions of the Swedish capitol. In six months the measure of his chest (a rude but the easiest test of improvement), had increased *four* inches, and in the course of the year his health was quite restored. At the time I saw him, several years after the event alluded to, he still retained his increased size, and his health was good.

The other instance alluded to is a physician, now Professor of Orthopœdic Surgery, based on the Movement-Cure. He also was far gone in consumption, when his attention being drawn to this method of cure, he resorted to it with the most complete success. Twelve years after, at the time of my acquaintance with him, he remained in good health.

It is important to remark, that it is not safe for this class of invalids to tamper with themselves without full directions.

### DEFORMITY OF THE CHEST, WITH PULMONARY DISEASE.

Pulmonary disease is often connected with deformity of the chest. The chest becomes flat on one side, loses its natural motions, the extent to which the air penetrates continually decreases, cough and expectoration supervene, and perhaps pain is felt on the well side, owing to its unnatural expansion in the attempt to receive the air excluded from the deformed side. Ordinary medical science furnishes no remedy, and the patient gradually sinks in a slow consumption. This condition may always be remedied by movements when taken in season, that is, before actual disorganization begins; and even afterward, the curative effects following restoration or even a partial restoration of the deformity, is often surprising.

The class comprises numerous cases, if we are to include the various grades of the affection usually unrecognized, even by respectable physicians. The following is an instance where the conditions above-described existed in an extreme degree:

CASE XI.—The Hon. Mr. B., M.C. from an eastern state, was recommended to my care by his physician, Prof. Peaslee. Seven years before he had a pleurisy. This was followed by the various evidences of pulmonary complaint, which increased steadily, till for the few previous months, he had become incapacitated for business. He had a severe cough : raised much thick, purulent matter ; was emaciated and weak.   An examination of the chest revealed that the left side had become much contracted in size, the shoulder being an inch and half lower, while the circumference of the affected side of the chest was an inch and half less than that of the opposite side.   Auscultation showed that very little air penetrated the affected side.

That the deformed and compressed side needed expansion, is the dictate of common sense.   That syrups, cordials, oils and hypophosphi es are incapable of effecting any such change, is equally obvious. Medical means are powerless for good, except to the extent they can supply *motion* and effectiveness to the respiratory function, because upon this fact hang the interests of vitality.   It is here plain that mechanical means are required to set the vital operations in healthy play, and that without such means the subject of the defect in question must gradually but certainly succumb.   The compression which hinders the action of the lungs, compels the blood to circulate with violent and fearful rapidity, wasting in the effort the little remaining vigor of the system, and thus increasing the disease.

As the process of cure went forward, the character of the matter raised was so changed that the yellow matter was replaced by mucus. The patient left at the end of two months to take his seat in the state legislature of which he was a member; the left shoulder being about *two inches* higher than before, and the circumference measure had increased nearly as much.   He has since called on me several times, a greatly improved man.

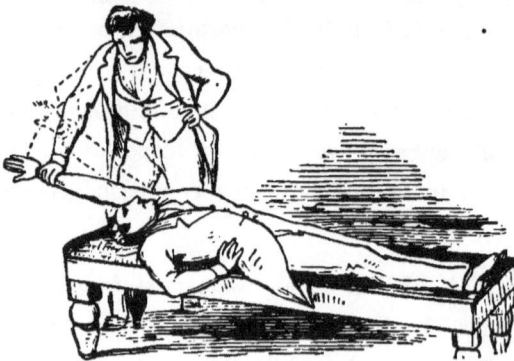

Fig. 15. — ARM RAISING —ACTIVE—Raising the ribs of one side of the chest, increasing the power of the muscles, and expanding that side of the lungs.

Similar instances might be given, if necessary, of cases with chest deformity, and immobility of one side of the chest, associated with disease of the liver, stomach, scrofula, &c., but those presented ought to be regarded as sufficient to illustrate the principle.

These cases of one-sided disease and deformity, afford an illustration as beautiful as complete of the *power* of the means employed, and supply, if needed, an irrefragable proof of its efficacy in other diseases.

## DEFORMITIES OF THE CHEST IN CHILDREN.

Medical skill has proposed scarcely any feasible way of restoring the deformed chest to its natural shape. Yet this is a very common affection and of very grave importance, rendering it impossible for the health to be otherwise than unsound, on account of the limit it imposes upon the respiratory function. Though the common practice of medicine is confessedly inadequate in these cases, the Movement-Cure is capable of demonstrating its entire adequacy, as may be seen in the following instance:

CASE XII.—Ida, youngest daughter of Mr. W., of this city, about nine years old, had a depression of the lower portion of the left side of the chest, the ribs being bent inward so much as to be quite flat, instead of having the natural convexity of the rounded, perfect form. The parents conjectured that the deformity was caused by a fall she received some years before. The measure of the deformed side of the chest, from centre of spine to centre of sternum, was clearly over an inch less than that of the opposite side.

The general health also suffered, so much so that her parents were under constant apprehension concerning her. The tongue was furred, the skin was of a dusky color, and apparently bloodless; she had frequent acute. attacks of indigestion; and she was always indisposed to engage in the recreations of children of her age, apparently from a consciousness of inadequate strength.

Fig. 16. —BENDING THE CHEST,
WITH DEEP INSPIRATION AND CLAP-
PING—ACTIVE AND PASSIVE—Expanding the deformed side of the
chest, and, by confining the other side, increasing the relative
amount of air inspired by the deformed side, enlarging the air cells
and restoring the natural shape.

In this case it was not only needful that the functions of the system be restored, but also that the mechanical impediment which hindered effective respiration, be removed before any permanent and satisfactory improvement of the health be possible. There must be an increase of the respiratory function, and the natural shape of the body, involving the bony frame-work, generally regarded as fixed, must be at least approximately restored. This will be regarded by those yet unacquainted with the Movement-Cure, as a difficult task. So it sometimes is, but it is by no means impossible, and often it is easy of accomplishment. The principles of *general* development necessarily includes that of development of *parts*. It only requires that the principles heretofore laid down be applied with tact and discrimination, and changes of shape are secured with a great degree of certainty.

In the case before us, abundant success attended the efforts made to accomplish the object above specified. The child rapidly increased in vigor and playfulness, and color again mantled the cheeks  Children, as might be presumed, are always more plastic than adults, and in this case the cartilaginous parts yielded with remarkable facility In six weeks the defective side was rounded out, though it did not acquire a perfectly natural shape; but the breathing power became quite as effective as that of the opposite side. The measure around the whole chest had increased *two and a half inches*, and I received the heartfelt thanks of the parents, who told me they had previously despaired of the ability of their child to approach maturity.

## PARALYSIS.

This affection consists of either defective power of motion, or its entire absence in some portion of the body. The disease is usually limited either to the lower extremities or to one side; and may exist in any degree, from slight weakness to loss of control by the will, of the part affected. It may come on stealthily or by a sudden shock. The *seat* of the disease is in those nerve centres of the spinal cord which communicate by means of nerve filaments with the affected part. The *nature* of this central disease is probably very different in different cases, though this is not indicated by the symptoms; hence the same amount of *apparent* disease is far from denoting equal facility of recovery.

There is no disease whose treatment by the usual methods is so unsatisfactory as this; and no doubt many cases have been rendered incurable by indiscreet attempts to *stimulate action*, when the development of a capacity therefor, should have been the object sought.

The following case will aid us in illustrating the success of the treatment by movements, and the principles involved in it:

CASE XIII.—Mrs. O., the wife of a physician in this state, aged about forty-two, had a sudden attack rendering her for a few days nearly senseless. She however rallied, but was confined helpless to her bed for several months. She afterwards became able to ride out, requiring a good deal of assistance in doing so. Her husband then brought her to this city for medical treatment, and by advice of Dr. Parker who examined her case in connection with Dr. Gilman, she came under my care. I found her left hand quite useless, but she was able to walk by the help of canes, requiring assistance in mounting stairs. She was also severely afflicted with uterine prolapsus.

The objects to be sought in this, and in most cases of this kind are two, viz: to *cause the mandates of the will to travel in the direction of the useless members ;* and to *remove congestion from, and restore healthy action in, the spinal cord.* These objects are to be sought in connection. The first is attained by increasing the freedom of the circulation of the blood, the temperature, and the general nutrition in the afflicted member, thus sustaining the conditions of vital power in both muscles and nerves together. In some cases this is enough, proving that the primary disease of the spinal centres had ceased to exist, and that it was only necessary to re-open the channel of communication with the central source of power.

But in most cases the spinal cord itself, where the power originates, must be subjected to healing influences. If there be congestion, it may be removed; if there has been destruction of the original sources of the nerve-power, the case is less favorable. Whether the one or the other condition exists, cannot be fully determined by the symptoms, but is proved only by the results of treatment.

The *processes* indicated, are first, such passive exercise of the affected member, as causes friction of the fibres which enter into its composition ; next, the bringing of the *will* to bear upon the same parts by means of such peculiar co-operation of an assistant, as increases greatly the degree of success. In this way the will is made to penetrate more freely the affected organ, and increased success is secured by the repeated efforts.

It is necessary to bring the spinal muscles into careful but energetic action. It is also quite important to perform such operations upon the fleshy parts over and contiguous to the spine, as shall result in drawing the congestion away from the cord.

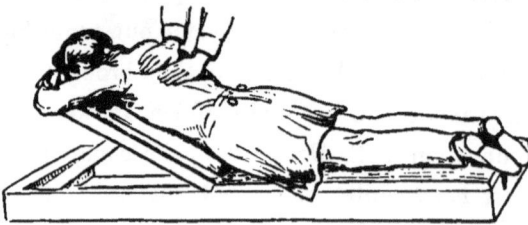

Fig 13. — FULLING THE BACK—PASSIVE — Increasing the amount of blood in the parts beneath the skin, withdrawing it from the cord, and inducing healthy action therein.

The paralytic invalid should bear in mind that his powers of elimination are feeble, and that his system retains much dead substance. If this is not an original cause of his disease, it is doubtless a consequence of it. He must rid himself of this by improving his respiration, increasing the action of his liver, more exposure to the air, more exertion, *improved diet*, and by all the hygienic means at his command.

The case at present referred to received treatment three weeks, with evident benefit,—carrying out the principles in practical detail, that are here only hinted at. Her husband, having learned the *why* as well as the *how* of the processes, thenceforward applied her treatment at home; with what advantage is indicated by the following extract of a letter I received from her about a month afterward:

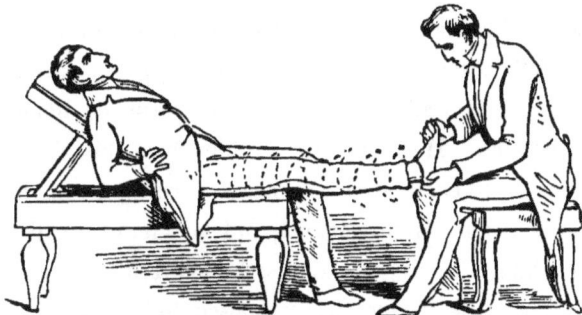

Fig. 14.—TWIST-SHAKING THE LEG—PASSIVE—Causing friction of the constituent fibres of the limb, exciting the nerves, increasing the amount of blood and heat in the part.

——— ———, OCTOBER 1st, 1860.

DR. TAYLOR:

DEAR SIR :—It is now about a month since I left my pleasant home in your house, and as I promised to give some account of myself after I left you, I hasten to do so. I am happy to say that my health prospers nicely, and that I gain more rapidly now than when with you. The treatment is pretty vigorously kept up, to which are added exercises not laid down in the books. I wish you could see how much I have improved in walking. I have walked to church two Sundays, morning and evening, and more than two blocks at once beside, positively without fatigue. . . . . . In short, I go about quite like a well person ; I am a miracle to myself and friends.

Now doctor, don't I give a satisfactory account of myself? and pray don't ever think again that there is no such thing as disinterested benevolence. I have nearly persuaded a patient of my husband's, with his assistance and hearty co-operation, to come to you. . . . . When I saw her last she had the blues dreadfully, and looked like all the woes. Poor thing ; she does not realize how speedily she might be brought up out of that, if she only knew how. . . . . . . .

Paralysis of the lower extremities also yields to the treatment, as will be seen by the following

CASE XIV.—Mr. II., from a western city, was brought here for treatment by movements. He was quite unable to walk, and had spasmodic twitchings of his legs; *complete incontinence of urine;* severe constipation, with lack of control of the discharges, numbness of the lower portion of the back and of the legs; soreness of the region of the liver, &c. He had also nearly lost his sense of sight. He received treatment with the following results: In three weeks he had gained control of his urinary discharges, and of the alvine, somewhat later. The natural sensations of the skin and flesh gradually increased, as well as his control of the muscles, till at the end of five weeks he was able to walk a little, when he returned home, travelling without assistance. He had full confidence of his ability to carry forward his cure to completion. Since returning home he has advised me of his continued improvement, and says he can walk better than I ever saw him. During his illness he had tried but too faithfully the ordinary routine of remedies proposed by the faculty. His disease had existed about eighteen months.

Paralysis generally requires a long and faithful application of the treatment, if the desired advantages would be gained, though a few are relieved speedily. It also demands most *careful* treatment, since it is so easy to overdo the weak parts, and excite anew the original diseased action, if under incompetent management. The non-recovery of a portion of cases, if the treatment be judicious, is to be attributed to the destruction of certain *nerve centres*, as vitalized and perhaps as organized parts. Since Science has, as yet, given no infallible test whereby the curability of a given case may be absolutely determined, it follows that a patient and persevering trial of the proper means, is the only consistent mode of discharging our duty to this class of invalids.

## SPINAL CURVATURE.

Deformity of the spine, of whatever kind or degree, is evidence of either present or previous imperfect muscular support of the spinal column, the consequence of imperfect nutrition of the spinal muscles. The spinal column consists of a series of separate bones or *vertebræ*, connected by elastic cartilage. It is therefore extremely flexible, being incapable without its muscular support, of maintaining itself in an erect position, much less of sustaining the mass whose weight rests upon it. Hence it appears, that the position of the spine depends entirely on the action of the muscles. While this mode of construction renders deformity possible in the case of weak muscles, it also facilitates recovery; for on the principles already explained, the nutrition of the muscles and even the bones, where this is faulty, may be increased almost at option.

A lateral single curvature, that is, one with but *one* convexity, does not present a difficult problem. One half of the body is weakest, including the leg, side of trunk and arm; nutrition should be increased in it, especially in those muscles whose contraction would cause the spine to become erect. The *cause* of muscular weakness referable to digestion must be reached, or all our endeavors however otherwise well directed, will be defeated. As the development of the defective side progresses, by processes similar to those by which deformity of the chest is corrected, the trunk resumes the erect position.

When the curvature is double, that is, having a *superior* convexity on one side, and an *inferior* one on the opposite side, the physician has a more difficult task before him; he has a complicated mechanico-physiological problem to solve. This problem has now been well studied, and a high degree of perfection has been achieved in the adaptation of means to the desired ends, securing oftentimes unlooked-for success by the treatment. The following is a good example, illustrating the principles necessary to bring into requisition in the treatment of curved spine:

CASE XV.—Mrs. C., a young married lady of this city, has always had a delicate constitution, and for several years very poor health. She had been much under medical treatment, availing herself of what is popularly supposed to be the best advice. She had the following symptoms: very pale and thin in flesh; backache; severe neuralgia; leucorrhœa; cough; had been treated much for pulmonary disease. The *right* shoulder was highest and projected backward, and to conceal the deformity she wore a thick pad on the *left* shoulder. In other words, the spinal column was curved to the right in the region of the chest, and also twisted backward; the transverse diameter of the chest deviating from that of the lower portion of the trunk by about the eighth of a circle. There was also fullness over the left hip, and a depression over the right, accompaniments of a lower curve to the *left* in the lumbar region.

To correct these curves, the left shoulder must be raised and developed; the arm strengthened, and fitted for habitual use; the left leg and left side strengthened and raised; the trunk untwisted; the ribs of the left side, which make too acute an angle with the spine, raised to the natural position; the plane of the pelvis, which dips to the left, must be raised to the horizontal; and the spinal column, now deviating to the right of the perpendicular in the thoracic region, and to the left in the lumbar, must be carried back to the erect position. These objects may in most cases be completely effected by the combined influence of the special strengthening processes of movements, applied to the weaker parts, and appropriate mechanical assistance. In recent and moderate cases, nothing more than well applied movements are required; but, as in the present instance, the patient frequently does not apply for aid till years after the curvature was contracted, something more than muscular weakness must now be overcome. Indeed, the *health* may have been quite restored by time, leaving the sad evidences of previous imperfect nutrition in the bones of the spine.

The shape of the vertebral bones have become altered in consequence of the unequal weight bearing on their two sides. They have become *thin* on one side—that of the most pressure—and thick on the opposite, being, in fact, wedge-shaped, the thick ends looking towards the convexity. The ribs also deviate from their natural position, being turned out, and making a greater angle with the column than natural on the *convex* side ; and turned down, in a less angle, on the *concave* side.

In my opinion. there is but one way possible to restore these long misshapen bones, and so to obliterate the deformity; this is, to reverse by artificial means the conditions by which the curve was formed. These bones were made thinner at one side by pressure, and pressure must be applied to the thick sides of the wedge-shaped bones to make them in turn thinner, and removed from the thin edges that they may thicken. This in fact is always done when an appropriate curvature movement is applied ; but for the special purpose of changing the shape of the bone and cartilage, it is necessary that the movements be increased in extent and greatly prolonged. For this purpose mechanical contrivances are requisite. The cut on the following page shows a mechanical device by which the objects here referred to are secured to the fullest extent.

A single movement by the apparatus affords a powerful exercise for the weak and low shoulder, elevating it with the ribs of that side, while the projecting shoulder is pressed in, as well as the lumbar fullness of the opposite side. In short, all the characteristics of the curvature are reversed, without pain to the patient. The padded bars admit of easy perpendicular adjustment, so that they may act on any horizontal plane, and so are readily adapted to any case of deformity. The whole apparatus may be fixed at any point to the post on which it slides, rendering it adapted to persons of any height, or to the sitting posture when this is preferred. The patient may leave the apparatus immediately after receiving the movement, or another

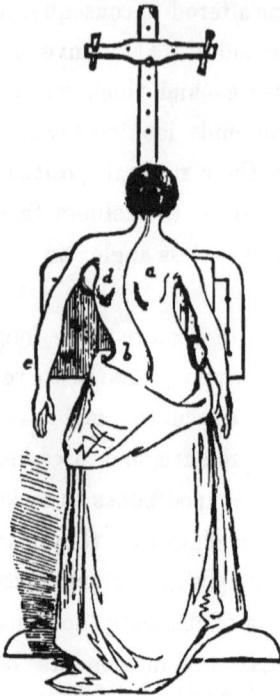

Fig. 17.—Showing position in apparatus of a case in which the upper curve is to the LEFT, lower one to the right.

Fig. 18. — Same as Fig. 17, but with the right hand grasping the lever, which being depressed, causes the padded bars to glide toward the centre, pressing in and obliterating the convexities, elevating the right shoulder, and reversing the position of the spine, in one motion.

padded bar, not shown in the cut, is turned under the elevated shoulder after relinquishing the hold upon the lever, sustaining it in its new position. The patient may remain in this position an indefinite length of time, amusing herself by reading or in any way fancy may dictate. The *twist* of the column, of which mention has been made, is counteracted by twisting, and other movements. But another apparatus is employed in the treatment, which combines *bending* with *twisting*. The use of the apparatus now referred to, renders the means of treatment complete, and is indispensable in the thorough management and rapid cure of these cases.

The object of this apparatus is to so prolong the combination of bending and twisting as to secure the moulding and shaping the bones of the trunk to their natural form. It consists of an attachment to the adjustible couch in common use in the Movement-room. To the edges of the flap are attached moveable loops, secured at any point to suit the curvature by thumb-screws. These loops receive uprights, which **may be two or three inches longer than the diameter of the body.**

The patient rests, sitting, reclining, or lying with the back against the flap. Very strong elastic bands pass round the body at the curvatures or points of greatest convexity, fastened to the uprights by hooks. These bands are elastic only two-thirds of their length, the continuation being of leather or some other inelastic substance. They are made tense around the patient, the ends being secured by the uprights of the opposite sides, and are tightened at pleasure by the patient by

Fig. 19.—COMBINATION COUCH—Bending and twisting the spine, moulding its vertebræ to the natural shape, and re-placing the ribs.

means of buckles. The elastic portion is so arranged that by contracting, the body will not only be *bent* but *twisted* also. The bending and twisting will not only quite obliterate the deformity, but may produce, for the time being, deformity in the opposite direction. The process is so gradual and gentle, as to be perfectly agreeable to the patient, who generally engages in some pleasant occupation.

A modification of the arrangement, also used, consists in continuing the bands at the upper side over a pulley at the top of the uprights, and attaching the ends to a roller under the centre of the flap. The roller is provided with a ratched wheel and a movable crank. When the crank is turned, one end of each band winds upon opposite sides of the roller, thus causing traction at the part over which it passes, *bending* the convex portion inwards, and by the same operation *twisting* the trunk upon its axis. This arrangement is somewhat preferable when the curvature is very old and rigid. In this case, the band passing over the most rigid portion at the right shoulder may be inelastic, serving as a fulcrum to those portions of the spine acted upon both by the elasticity of the other bands, and by force derived from the crank. This arrangement may be adjusted in regard to the points to which the pressure and twisting should be applied, the degree of elasticity and of power required, the position of the patient, whether sitting, lying or intermediate positions, with the greatest case. It also supplies an almost unlimited amount of power. It does not uncomfortably press, cramp or strain the most delicate person; and such frequently indulge in a luxuriant nap while in position. In short it is difficult to conceive any practicable purpose which this arrangement does not completely fulfill. The lady whose case has been stated in this connection, occupied the couch perseveringly, two hours every forenoon, and the same time every afternoon, and in *four weeks* she discontinued as unnecessary the pad of the left shoulder, which previously concealed her deformity, and her dresses no longer fitted the altered shape. In evidence of the expanding effect of the apparatus, and the throwing outwards of the depressed and deformed ribs, the fact may be stated that at the end of six weeks' treatment, her chest measure had increased three and a half inches. In the meantime the neuralgia was cured, the cough vanished, color came again to the cheeks, and vivacity of spirits, and all the elements of good health were restored to our patient.

It is indispensable for the successful treatment of deformities that mechanical and physiological treatment be combined; indeed either method alone, if not quite useless, will only partially secure the desired object. Braces and supports weaken the parts, and ultimately increase the deformity. In a former patient of Tamplin recently examined by me, the body was twisted to the extent of a quarter of a circle—an effect evidently *produced* in great part, by the dragging weight of the iron arrangement intended for restoration.

www.ingramcontent.com/pod-product-compliance
Lightning Source LLC
Chambersburg PA
CBHW021630270326
41931CB00008B/955